© 1992 SCOTT E. SUTTON

# The Kuekumber Kids Meet
™

# THE SHEIK OF SHAPES

© 1992 SCOTT E. SUTTON

Written and Illustrated
by
SCOTT E. SUTTON

Published by

SUTTON PUBLICATIONS, INC.

EDUCATION
THROUGH
IMAGINATION

TM

14252 CULVER DRIVE, SUITE A-644
IRVINE, CALIF., 92714

First Edition
ISBN 0-9617199-9-0

Copyright © 1992 by
Scott E. Sutton
All rights reserved

Special thanks to the following for all their help, encouragement and inspiration:

My wife Susie, her parents, my parents, all my teacher friends,
Applied Scholastics, the American writers Dr. Seuss and L. Ron Hubbard,
and my art teacher Tom Morgan.

Printed in Thailand

# THE SHEIK
# OF SHAPES

# Meet the Kuekumbers

This is Kirky Kuekumber.
He'll be six soon.
When he sees an adventure
He's out the door ... zoom!

This is Katie Kuekumber
Who is now age four.
She asks lots of questions.
And then she asks more!

This here is Krumby,
The Kuekumber kids' dog.
He's a really great pooch
But he eats like a hog.

This is Kim,
The Kuekumber kids' mother.
She takes care of Katie
And Katie's big brother.

This is Karl,
The Kuekumber kids' pop.
He likes to build things
In his little workshop.

This is Kirky's
Best friend, Lance,
Who likes to wear
Big, baggy pants.

© 1992 SCOTT E. SUTTON

Katie and Kirky were in Lance's front yard.
And to see them you'd think they were working so hard.
Lance had some building blocks made out of wood
And the kids were stacking them however they could.

They made buildings and bridges and towers, too.
They built a whole city before they were through.
Lance made a building with a tower on top.
Katie made a tower that fell down with a PLOP!

2

"Weeee, this is fun!" Katie Kuekumber said,
Even though some blocks fell down on her head.
Then Krumby started barking into the air.
"Why's he barking?" asked Kirky. "Is something up there?"

The kids stopped building and looked to the sky
Where they saw something weird flying nearby.
"It looks like a guy on a carpet to me,"
Said Lance, who was trying his best to see.

It swooped down low while making a pass.
Then the flying carpet floated down onto the grass.
On the carpet was a strange, short little fellow
Wearing long, flowing robes of red, white, and yellow.

"Greetings, children!" he said with a wave of his hand.

He looked like a king from a far-off land.

He stepped off his carpet and over he came

To where the kids were. They asked, "What's your name?"

"My name is Razoolee Banzaleenee Parmizong,
But just call me Sheik 'cause that name's too long.
And I've flown from my kingdom far, far away
To teach you all about shapes today."

"You mean," asked Lance, "like a box or a ring?"
"Yes, shapes!" said Sheik. "They're the outlines of things.
And everything in the world has a shape, you see,
From a bicycle wheel to the leaves on a tree."

"Now, to help me teach you without a mistake
I've brought my trusty assistant, Shape Snake."
Shape Snake wriggled over without a care
And brought Sheik his pencil that could write in the air.

9

"Now, the first thing I'll draw with this pencil of mine
Is the thing you build shapes with. It's called a **LINE**.
You make a straight mark from point A to point B.
It's as simple as that! A **LINE**, don't you see?"

10

"Now if you bend this line here in the middle,
Then the line becomes **CURVED**, a lot or a little.
If the **LINE** goes up and down alot
A **ZIG ZAG LINE** is what you've got!"

"So what do you call it when two lines meet?"
Asked Katie Kuekumber as she jumped to her feet.
"That's called an **ANGLE**," said Sheik as he drew.
"With **ANGLES** there're all kinds of shapes you can do."

The Sheik drew four lines, made a shape in the air,
And said to the kids, "This shape is a **SQUARE**.
It's sides are the same.  It has only four.
And a **SQUARE** has four **ANGLES**, just four and no more."

"This shape is a **RECTANGLE**. It has four **ANGLES**, too.
But it's different than a **SQUARE** and here's what you do.
Draw two sides long and two sides short.
Now you have a shape of the **RECTANGLE** sort."

"The next shape," said Sheik, "is a **TRIANGLE**, you see.
If you count all the **ANGLES** it adds up to three!"
"It has three sides, too," said Kirky Kuekumber.
"You're right," said the Sheik. "Three is the right number."

15

Sheik drew two **TRIANGLES** the exact same size

And built a new shape.  The kids were surprised.

"Take two **TRIANGLES**," said he, "put them back to back.

And you've got a **DIAMOND** shape, just like that!"

16

"Now, this next shape I'm drawing is perfectly round."
The Sheik started spinning as he drew near the ground.
"From the middle to the edge is always the same.
It's made from a **CURVED LINE** and **CIRCLE**'s its name."

Then Sheik drew a **CIRCLE**, pushed the sides in a little.
So it wasn't the same distance from the sides to the middle.
"That's an **OVAL**," said Lance. "I know that shape.
That is the shape of an egg or a grape."

"Now I've shown you," said the Sheik, "shapes that are flat.
But there are shapes that stand out, like a ball or a bat.
These shapes are long and tall and wide,
And you can look at them from every side."

"First is a square box.  It is called a **CUBE**.

It looks like 6 **SQUARES** that together are glued.

But if you have a box shape whose sides aren't the same

It's a **RECTANGLE BOX** and that is it's name."

"Now **PYRAMIDS** are made of **TRIANGLES**, you see.
They either have four sides, or they have three.
On the four-sided one the bottom's a **SQUARE**.
Lift a three-sided one and a **TRIANGLE**'s there."

21

With his pencil the Sheik made a round shape appear.
It looked like a ball but he called it a **SPHERE**.
The Shape Snake crawled over and gave Sheik a grape.
"Thissss," hissed the Snake, "is an **OVAL** shape."

"How 'bout a glass?" asked Katie. "What shape would that be?"
"A **CYLINDER**," said Sheik, and he drew one quickly.
"And this round-bottomed shape is very well known.
You can put ice cream in it, it's called a **CONE**."

"Now," said the Sheik, "to my flying carpet, please.
We're going to fly up high above all the trees.
And what I want you kids to do
Is spot the shapes I've been teaching you."

They climbed on the carpet and flew to the sky.

The kids named the shapes as they saw them go by.

"That building's a **CUBE**!" Lance said right away.

"And the moon is a **SPHERE**," Kirky did say.

"That roof's a **SQUARE**," yelled Katie, with a happy face.
The kids spotted shapes all over the place:
**CYLINDERS**, **PYRAMIDS**, and **OVALS**, too,
And **CUBES**, and **CONES** before they were through.

26

"Excellent!" said Sheik. "Yessss!" hissed Shape Snake.
"You've named shapes," said Sheik, "without a mistake.
But now I can see by the setting sun
It's time we get going, our work here is done."

So the Sheik and Shape Snake flew the kids home,
Dropped them off, said goodbye and flew off alone.
Back to their kingdom far across the sea –
The kingdom of shapes – now where could that be?